SCORE READING

BOOK I

ORCHESTRATION

BY

ROGER FISKE

MUSIC DEPARTMENT

OXFORD UNIVERSITY PRESS

WALTON STREET · OXFORD OX2 6DP
200 MADISON AVENUE · NEW YORK N.Y. 10016

© *Oxford University Press 1958*
Fourteenth Impression 1987

ISBN 0 19 321301 X

Printed in Great Britain by
J. W. Arrowsmith Ltd., Bristol

INTRODUCTION

THESE books are intended primarily for students of all ages who are interested in learning how to read an orchestral score. They are for use in conjunction with recordings of the works listed on p. 72. This first book consists of some eighteen movements printed in full score, and chosen for the ease with which they can be followed and for the clarity with which they exhibit the principal instruments of the orchestra. These movements have been reproduced from a variety of sources, an economy that has the advantage of introducing the reader to most of the different ways of setting out a score that he may meet later. In schools, the need for such a book will be recognized by any teacher who, when requesting the purchase of an expensive set of miniature scores, has found himself concealing the fact that only one movement in the work is going to be of any use. For beginners in score-reading only slow movements are practicable, and these must be for as few instruments as possible. Such movements are not too easy to find, and, when found, are difficult to grade. Experience may show that they are not properly graded in this book; the user is of course free to amend the order of the material.

The concentration of slow-moving pieces in the first book may seem unattractive on the grounds that young people especially prefer quick gay music and lack the patience for anything else; but in fact it is one of the virtues of score-reading that it helps one to concentrate on music that would otherwise seem uninteresting. After all, quick movements can be played as an antidote to score-reading. Following music can reveal elements of orchestration and form, and thereby increase appreciation of a piece of music and admiration for its composer. There is something to be said for dispelling the illusion that music is a mystery which only those divinely inspired can hope to create. Writing music is in fact a craft like making a bookcase or building a radio, and the composer's score is a practical affair, planned in detail for a specific purpose.

My choice of pieces has inevitably been limited by such factors as length and the laws of copyright. However suitable in other respects, a movement taking twenty or more pages could not justify its inclusion, and modern copyright works have been omitted in order to keep the cost of the book down as far as possible.

The first book aims to teach something of orchestration through score-reading, the second something of musical form. At the end of each book there is an index of the music reproduced in the pages that follow.

HINTS FOR BEGINNERS

1. Look at the tempo mark before the music begins.

   This will give you a rough idea how quickly your eye has got to scan the music. (For the meanings of tempo marks, see page 70.)

2. Follow the first violin part; more often than not the first violins have the tune. But if the first violins have rests or are accompanying, glance up at the woodwind and brass staves until you find a line that looks interesting.

3. Look out for characteristic rhythmic groups and repeated melodic phrases. Most music is based on frequent repetition of a rhythmic pattern, or of a melodic shape, or both; sometimes only three or four notes in length. Look for a repeated pattern before the music starts. There will probably be a new set of rhythms and shapes called 'the second subject', but these will not be so important as those in the first few bars.

4. In between the tunes there may be passages in which the violins play too quickly for you to follow the individual notes. Either follow the rhythm in the bass, or skip on to the first point where familiar rhythmic patterns reappear and wait for the music to catch you up.

5. When you are able to follow the melody in a piece, start looking at what the other instruments are playing. You will soon develop an instinct for distinguishing the interesting staves from the purely routine ones. Do not expect ever to be able to follow every stave at once; very few people can do this, and score-reading can be enjoyed at a much less advanced level. But you will often need to keep an eye on two staves at once.

FOLLOWING A MELODY

LOOK for patterns of quicker notes, and learn to take them in not individually but as a group,

e.g. the rising quavers (eighth-notes) in bars 1, 3, &c., below.

Look for wide leaps in the melody, and learn to associate the *look* of them with the *sound*,

e.g. in bar 4 below.

Conversely, look for passages in which successive notes are next door to each other, and leaps kept to a minimum,

e.g. in bars 5 to 7 below.

Repeated notes can be an anchor for the eye,

e.g. in bars 2 and 4.

This is a minuet that Bach wrote for his wife when he was teaching her to play the harpsichord. It is often given to beginners on the piano and is probably familiar to you. Each half is played twice. The thick line after bar 16 with dots before it is a 'repeat' mark, and means 'Play bars 1–16 again.' Similarly at the end, but here you are to go back to the thick line with the dots *after* it at the start of bar 17. You will only be following the melody, that is, the top line of each pair, but you will need to glance at the lower line in bar 8.

J. S. Bach

Notice the rhythmic patterns in this tune. Bars 3–4 have the same rhythm as bars 1–2; bars 5, 6, and 7 all have the same rhythm, and it is the same rhythm as bar 1. Once you have taken in the rhythm of the first bar, you could follow this music without thinking about the rise and fall of the melody at all; you could just follow the pattern of crotchets and quavers (quarter-notes and eighth-notes). In fact you will find yourself giving about equal attention to melodic outline and rhythmic pattern.

FOLLOWING TWO STAVES

CONSIDER these two tunes; the first is Handel's so-called 'Largo' (really a song from an opera about Xerxes) and the second Bach's so-called 'Air on the G string' (really the Air from his Suite No. 3 for orchestra). Both are slow.

You will not hear the second note in these tunes; it is 'tied' to the first, of which it is a continuation in the next bar. This apart, you will find such music hard to follow from the tune alone because you can hear sounds that you cannot see. With the bass accompaniment printed they are much easier to follow; your eye takes in that the tune starts with a long note and while this is going on you watch the bass. This gives you the rhythm that you hear. When the tune starts moving (and your ear will tell you when this happens, even if your eye does not) you cease following the bass and watch the tune instead.

Handel

J. S. Bach

The Bach Air is printed in full on p. 56. It is harder to follow than the earlier pieces in this book.

RHYTHMIC CELLS

CONSIDER these two tunes:

They look very alike. Each starts with the same rising three-note phrase (which is repeated) and ends with a much longer phrase. Yet they do not sound like each other, and no one has ever accused Brahms of cribbing the top tune, which Grieg had written four years earlier in his *Peer Gynt* music. The differences are instructive:

1. The Brahms tune goes quicker, and your eye will have to follow quicker.
2. The Grieg is for strings and the Brahms for piano, which means that the *quality* of the sound is quite different.
3. The Grieg has chords which for the most part are in the same rhythm as the tune (see p. 12); thus the tune is all that matters for the inexpert score-reader. But the Brahms has a running accompaniment that transforms the music:

You will not be able to follow this running part very easily, and there is no need to do so. What matters is that you should *see* that there is a running part and *hear* it against the tune. Even though you are only following the tune with close attention, it is easier to do so if you have the running part in front of you as well, for you then have a complete picture of the whole—which in this case is very different from the part, the tune alone. Score-reading is usually easier when all the staves are in view, not harder as you might expect.

Now let us take each of the above tunes in turn and see to what use each composer puts his rhythmic patterns. (There will be no space for similar analyses of the other music in this book, and no need, for much the same principles can be applied to most music.)

Grieg (see p. 12) repeats the same pattern until near the end of his 'Death of Åse'; each four bars has a three-note phrase repeated, followed by a longer phrase with a pair of quicker notes in the middle. You should be able to recognize these units almost from the start, and from then on following will be easy. Notice that after eight bars Grieg repeats exactly the same music five notes (a fifth) higher. Nothing could be simpler. He then returns to his first eight bars but much louder.

Brahms's Rhapsody in B minor is a far more complex piece of music, much of it too hard for beginners to follow. After some stormy music the tune printed on the left provides contrast. If you can hear this piece played on a piano, or coming from gramophone or radio, you will find that Brahms also repeats his four bars higher up. Then after some more stormy music there is a 'middle section' starting like this:

Follow the big notes in the treble stave and do not give much attention to the others. Your ear may not tell you anything about this tune except that it is a good one. Your eye can come to the help of your ear and show you that in fact it is based on bars 3 and 4 of the tune from this Rhapsody that we discussed first; Brahms has disguised it by putting it in the major with a smoother accompaniment. Composers often link their ideas in this way to give their music unity, and although the listener may not be aware of the link, he is probably aware of the unity; he 'feels' it without knowing how it has been achieved. One of the great pleasures of score-reading is the discovery of how music is put together; eye and ear can be complementary to each other.

THE STRINGS

Music for strings is usually printed on four staves. The violins are divided into two sections, 'Firsts' and 'Seconds', and they have a line each. The third stave down is for the violas, which look just like violins except that they are a little larger and play a little lower. The fourth stave is usually shared by cellos and double-basses. The double-basses usually play the same notes as the cellos but the sound comes out an octave lower.

Both the pieces for string orchestra that follow were written for the stage, and both are connected with old age. 'The Death of Åse' is played as a background to a long speech by Peer Gynt boasting to his old mother of all his wonderful deeds, most of them his invention; he is quite unaware that she is dying. The Adagietto by Bizet was written for Alphonse Daudet's play *L'Arlésienne*, and accompanies the meeting of an old shepherd and an old farmer's wife; they have not seen each other since they were young and in love with each other. The music exactly reflects this sentimental but very moving occasion.

ADAGIETTO FROM *L'ARLÉSIENNE*, BY BIZET

In 'The Death of Åse' on the next page, the double basses do not play the same notes as the cellos, and so need a stave of their own. From bar 17, first and second violins and violas are *divisi* or 'divided'; some play the upper notes and some the lower.

♩ = 50 : at a speed of 50 crotchets (quarter-notes) a minute.

con sordini: with mutes.

∨ (bar 5 &c): up bows. ⊓ (bar 9, &c.): down bows. Usually composers leave such details to the players to work out.

THE DEATH OF ASE FROM *PEER GYNT*, BY GRIEG

THE WOODWIND

You will now meet the principal wind instruments in the orchestra in the order in which they are to be found in the conductor's score.

The woodwind are arranged thus: flutes, oboes, clarinets, bassoons.

At first you will find it easier to follow music for these woodwind instruments that can be accommodated on four or five staves, that is, chamber music. Mozart (1756–91) wrote a very large quantity of chamber music for all kinds of combinations, most of it for his friends to play rather than for publication. Below is the slow movement of a quartet for flute and strings (violin, viola, and cello). It has no definite end, but leads straight into the finale (not printed here; it is fast-moving and you will enjoy it more easily without following). The strings play *pizzicato* throughout; that is, the players pluck the strings with their fingers instead of using their bows, and the result sounds rather like a guitar accompaniment to a serenade.

There are a number of 'appoggiaturas' in the flute part—small notes placed immediately in front of normal-sized notes (see, for example, bar 2). These small notes, which are especially common in eighteenth-century music, come *on* the beat and steal some of the value of the next note, but it is not quite certain if Mozart wanted ♩♩ in bar 2 to be played ♩ ♩ or ♪♩..

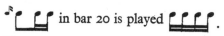 in bar 20 is played ♫♫.

Flute

Flauto traverso means cross flute, as opposed to the recorder type.

SLOW MOVEMENT, FLUTE QUARTET IN D K. 285, BY MOZART

[14]

SLOW MOVEMENT, OBOE QUARTET IN F
K. 370, BY MOZART

Oboe

The music opposite is the slow movement of Mozart's Oboe Quartet (K.370). The three stringed instruments are the same as for the Flute Quartet, but here the first violin has a much more interesting part and you will have to give almost as much attention to the second stave as to the first. In the third bar of the last system Mozart put a pause on the first note, and this means that he expected the oboe player to extemporize a short cadenza, that is, make up a few runs and flourishes (but not brilliant ones, for this is a slow movement) finishing with the trill you see in the second half of the bar; this trill is a signal to the string players that the oboist is finishing the cadenza, and they get ready to join in at the next bar. Unfortunately many oboists think that the pause mark means a pause and nothing else—which it would in music written in the nineteenth and twentieth centuries—and they do not play a cadenza. But in eighteenth-century music the sign ⌒ is nearly always an invitation to a soloist (or singer) to show off some technical tricks; it is the accompanying instruments that pause.

Each of the main woodwind instruments has a near relation, either larger or smaller than itself. Flautists, for instance, must be able to play the half-length flute called the piccolo, and some clarinettists and bassoon players can also perform on the bass clarinet and the double bassoon respectively. Perhaps more important, in the sense that it is more often given solos, is the large oboe illogically known as the cor anglais or English horn. This can play a fifth lower than the oboe. Players naturally wish to use the same fingering for both instruments, so that when they see the note Middle C they automatically cover the same holes. On the longer cor anglais this fingering naturally produces a lower note; *the sounds are always a fifth lower than the written notes*. Conversely the composer has to write the cor anglais part a fifth higher than the sounds he wants. In this book there is only one short tune for the cor anglais, and it will be found on page 29. The famous cor anglais solo in Dvořák's 'New World' Symphony is given in Book II.

TRILLS: Among the useful shorthand signs that composers employ is *tr* (trill); there are two examples in the oboe part opposite and one in the flute part on the previous page. The player oscillates very rapidly between the note printed and the one above. This flute part also contains several written-out 'turns' in the bottom line. There is a shorthand sign for turns: ∽. For examples see the clarinet part on pages 19–20; also page 63 and 65–9.

The clarinet is another 'transposing' instrument; the sounds it makes are not those that the composer writes. This presents no particular difficulty to anyone following a clarinet stave on its own, but score-readers who reach the point of trying to hear the music on the next page in their heads, or even just the top two staves, need to understand the principles of transposing instruments.

Clarinet

The clarinet was the last of the main woodwind instruments to become established in the orchestra, and Mozart was the first great composer to realize its possibilities. In his day clarinettists found music with many sharps and flats extremely hard to play, and they got round this difficulty by having three instruments of slightly different length and choosing the one best suited to the key of the music. Or rather the composer chose for them, and indicated for which clarinet he was writing. The clarinet in C behaved normally; it was *not* a transposing instrument. It was seldom used unless the music itself was in C major. For flat keys players used the slightly longer clarinet in B flat; being longer, the notes came out lower, a tone lower. The player presses down his fingers to play example A and the sound is example B.

Thus, while all the other players have to cope with two flats in the key signature, the clarinettist has no flats at all. In a very flat key he will find *some* flats in his key signature, but always two less than everyone else. For sharp keys he uses a still longer instrument, the clarinet in A, sounding a tone and a half lower. The composer writes C, and the sound is A. Thus the music on the next page will sound:

Nowadays the clarinet in C is no longer used, though you will still find it in many scores by Mozart and Beethoven. Modern clarinets are easier to play than those Mozart knew, sharps and flats are not so hard, and players can manage any parts in C that they meet with on the B flat instrument, transposing in their heads as they go. But it is still a convenience to them to have the two instruments in B flat and A, however inconvenient for score-readers. Roughly speaking:

The B flat clarinet sounds one note lower (i.e. a tone).

The A clarinet sounds two notes lower (more accurately, a tone and a half or a minor third lower).

Mozart wrote his great clarinet works, the Quintet and the Concerto, at the end of his life for a somewhat disreputable friend called Anton Stadler. In the slow movement of the quintet, printed opposite, the clarinet has the main tune, but from bar 20 the first violin shares the interest until the return of the main tune. Your eye will seldom be distracted from the top two staves except in the last three bars where viola and cello have the interest for a few brief moments.[1]

Notice how Mozart makes use of the low notes on the clarinet, what is called the 'chalumeau' register. The clarinet can go much lower than the flute or oboe.

SLOW MOVEMENT OF MOZART'S CLARINET QUINTET

[1] The third and fourth movements of this quintet will be found in Book II.

We have now reached the last of the major woodwind instruments, the bassoon, and our first orchestral score. The music on the next page is by the French composer, Georges Bizet (1838–75), and it is played as an entr'acte in one of the intervals in his opera *Carmen*. It is usually known as *Les Dragons d'Alcala*, Alcala being a part of Seville in Spain, and the 'Dragons' of course being dragoons.

Bassoon

There are two ways of printing a full score. The easiest for both score-reader and conductor is to have a stave for every instrument on every page; thus any particular instrumental part always occupies the same position on the page. You will find this method adopted in the admirable Penguin scores. Its disadvantage is that it wastes paper, for in most music there are long stretches during which some of the instruments do not play, and thus many pages are full of staves consisting entirely of rests. Hence the alternative method, in which instruments have no staves when they are not playing. Thus several 'systems' as they are called can be accommodated on a single page, for instance three on the next page. With scores printed like this you need to keep a sharp eye on the names of the instruments. On the first page, follow the bassoon, and in the final bars notice how all the woodwind share the tune in turn. Notice that at the start the strings are told to pluck *pizzicato*, but at the top of page 25 (last bar) this direction is cancelled by the word *arco*, 'with the bow'. The 'Kleine Trommel' that plays the fifth stave down at the start is a side-drum.

Most scores give the names of instruments in Italian, but some use German or French. The following list may be found helpful:

	ITALIAN	GERMAN	FRENCH
Flute	Flauto	Flöte	Flûte
Piccolo	Piccolo	Kleine Flöte	Petite flûte
Oboe	Oboe	Hoboe	Hautbois
Cor anglais	Corno inglese	Englisches Horn	Cor anglais
Clarinet	Clarinetto	Klarinette	Clarinette
Bassoon	Fagotto	Fagott	Basson
Horn	Corno	Horn	Cor
Trumpet	Tromba	Trompete	Trompette
Trombone	Trombone	Posaune	Trombone
Timpani (Timps)	Timpani	Pauken	Timbales
Viola	Viola	Bratsche	Alto

The viola is the only stringed instrument about whose name there can be doubt. Note that in German B means B flat, and As, Des, and Es A flat, D flat, and E flat respectively.

LES DRAGONS D'ALCALA FROM *CARMEN*, BY BIZET

Here is another entr'acte from *Carmen*. The previous one contrasted bassoon and clarinet; this Intermezzo contrasts flute and clarinet. But only the first part of it is given here as the end is rather hard to follow. The cor anglais is a large oboe with a deeper voice (see p. 16); the notes sound a fifth lower than written.

THE SLOW MOVEMENT OF BACH'S SECOND
BRANDENBURG CONCERTO (see p. 32).

When Bach wrote his six concertos for the Elector of Brandenburg, and this was about 1720, almost all music had a 'continuo' part, that is, a bass line played by a cello (or viola da gamba) 'e Cembalo', which means 'and harpsichord'; the harpsichordist was expected to invent chords above this bass line as he went along, and so fill in the harmony. Usually the composer put figures under some of the notes of the bass line and these were a kind of shorthand indication of what chords he should play. In the slow movement reproduced on the two preceding pages, you will not need to pay much attention to the continuo part, which walks in even quavers (eighth-notes) almost throughout the movement. But the other three staves will all need close attention. The subject to be discussed by the three instruments is played first by the violin (first nine notes) and the rhythm of this subject permeates the whole piece. But notice also a tiny sighing phrase of three notes that comes next in the violin part:

BRASS INSTRUMENTS

THE brass instruments are arranged in orchestral scores in this order:

> Horns (usually 4),
> Trumpets (usually 2),
> Trombones (usually 3),
> Tuba (usually 1, and only in music of the last hundred years or so).

In modern music there are often more horns and trumpets than are suggested here. Trombones and tuba will not concern us in this book; they seldom have solos, and are used mostly for adding weight and excitement to climaxes, though trombones are also effective playing very soft chords.

Horns and trumpets are transposing instruments of a similar type. In the time of Bach, Mozart, Beethoven, and Schubert (that is, until about 1830), horns and trumpets had no valves. The 'natural' horn was a simple brass tube, coiled for convenience; by changing his *embouchure* or lip pressure the player could theoretically produce the following notes:

This is called the harmonic series. With exceptions to be dealt with later, these are the only notes that composers before 1830 ever *wrote* for the instrument, and it will be noticed that a tune demanding any scale figures is only possible at the top of the compass.

Now I have assumed that the tube is of such a length that the instrument produces the harmonic series on C. In fact the players had a number of 'crooks', extra lengths of tube which they could insert into the main tube to make it longer. The ones most commonly used were the B flat alto, A, G, F, E flat, D, C, and B flat basso. Each was longer than the one before and produced a lower harmonic series. *But the composer always wrote the notes of the harmonic series in C major, as above.* The *sounds* that resulted depended on the crook.

Let us take a common progression for horns in the eighteenth century:

with that music in front of him, a player who put in some of the crooks listed above in succession would produce the following sounds:

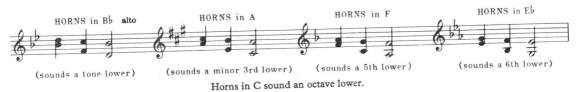

HORNS in B♭ alto   HORNS in A   HORNS in F   HORNS in E♭

(sounds a tone lower)   (sounds a minor 3rd lower)   (sounds a 5th lower)   (sounds a 6th lower)

Horns in C sound an octave lower.

The trouble was that it was next to impossible to change a crook during a movement, for if you did not push it in exactly the right amount the instrument would be out of tune; and you cannot tune a horn in the middle of a symphony. Thus as soon as the music left the main key, horns and trumpets became practically useless.

About 1830 valve horns started coming into use. Valve horns did not need crooks, for they had them built in. Each of the three valves (they are worked by knobs on some horns, levers on others) added a certain length of tubing; any two valves could be pressed down, or all three. Thus all necessary harmonic series became available, and the instrument could be used chromatically throughout the greater part of its compass. For nearly a hundred years, all (or almost all) horn parts have been written as though the horn had an F crook built into it, regardless of the key of the music. The composer writes C when he wants the F below. Thus the sound is a fifth lower than the notes. But in classical music you have to look at the first page of the score to find out what key the horns are in.

The seventh note of the harmonic series happens to be rather out of tune, and before about 1800 composers seldom used it. Thereafter players found that they could get it in tune by putting their hand inside the bell of the horn; in fact they learnt to fill in all the notes between the sixth and eighth harmonic by this means, and some other chromatic notes as well, but these notes never had such a good tone as the normal kind.

For the last hundred years or so, trumpet parts have been written in C or B flat (occasionally A and F), but in classical times they had as many crooks as the horn. Since the invention of valves, horns and trumpets have had much more interesting music to play.

Horn

Below is the Nocturne from Mendelssohn's incidental music for Shakespeare's *A Midsummer Night's Dream*, written for a production in Berlin of about 1840. This piece is played as the curtain falls on Act III after Puck has put all the lovers to sleep in the wood near Athens by squeezing juice in their eyes. The curtain rises again on the next scene some fourteen bars from the end, where the mood of the music changes to suggest Titania, Bottom, and attendant Fairies entering and discovering the sleeping lovers.

Though valve horns were coming into use, Mendelssohn wrote for the 'natural' horn, and asks for the rather uncommon crook in E.

NOCTURNE FROM *A MIDSUMMER NIGHT'S DREAM*, BY MENDELSSOHN

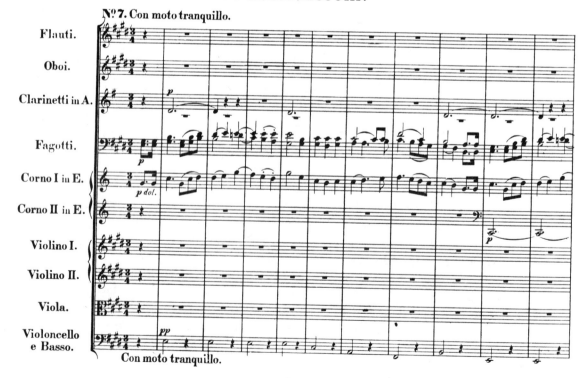

(Hier öffnet sich die Laube wieder und man sieht Titania und Zettel mit den Elfen.)

(Oberon im Hintergrunde verborgen.)

(The bower opens again; it is to see Titania and Bottom with the elves.)

(Oberon behind, unseen.)

HAYDN'S TRUMPET CONCERTO

At the time of Bach and Handel, trumpeters cultivated the top notes on their instruments, being able to reach the high C two notes higher in the harmonic series than the notes used by later composers (see p. 32). You can hear high trumpet parts in 'The Trumpet shall Sound' in Handel's *Messiah*, and in the Finale of the second Brandenburg Concerto by Bach, of which the slow movement is given on

p. 30. About 1750 the art of playing these high notes (which are very exciting to hear) deteriorated, and it is still not clear why this happened. The horn was being used more and more in orchestras, and possibly trumpet players transferred to the horn because horn parts were so much less chancy; playing Bach's trumpet parts is not only very hard work, but also a great

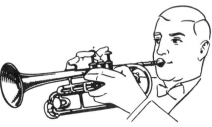

Trumpet

strain on the player, who lives in constant terror of 'breaking' on a high note even today.

Whatever the reason, trumpet and horn parts became tame dull affairs from about 1750 onwards. But not everyone was satisfied with the situation, and about 1795 someone in Vienna invented a trumpet with keys like a woodwind instrument and holes in the tube. It was not a very successful invention and never came into general use, but during its brief career the inventor persuaded Haydn (1732–1809) to write a concerto for it. This concerto is quite unplayable on the natural trumpet, and had to wait for the invention of the valve trumpet before it could be much played anywhere. The slow movement is reproduced in the pages that follow.

Notice that in a concerto the solo instrument's stave is placed between the wind and the strings; in other words, just above the first violin. This is a convenience, for most of the time in this movement these are the only two staves you will need to watch. The violins have the main tune (first eight bars), and then this is repeated by the trumpet.

The first and last movements of this concerto are in E flat, the key of the work as a whole, and Haydn naturally wrote for the trumpet in E flat, and though his slow movement is in a different key (A flat), he did not bother his soloist by asking for any adjustment. (Perhaps none was possible; it may well be that this instrument was built in E flat.) Unlike horn parts, trumpet parts quite often transpose upwards, and in this case the trumpet notes sound a third *higher* than written. Thus bars 9–10 on the trumpet sound at the same pitch as bars 1–2 on the first violins. In bars 21–24 and 45–46 trumpet and first violins are 'doubling', that is, producing the same notes.

FULL SCORE

So far in this book you have not been asked to give attention to more than two or three staves in any one piece. In the full scores of two well-known pieces from Schubert's incidental music for a play called *Rosamunde* nearly every stave is of importance at one time or another. In the Entr'acte you may be able to manage with the first violin stave alone as far as the second double bar; in the section headed 'Minore I' (this merely means a contrasted piece of music in the minor key) clarinet, oboe, and flute staves should hold your interest; the strings can be ignored. But when you know this music better you should notice that there is more in orchestration than just contrasting one solo instrument with another; there are also infinite possibilities in 'doubling'. Thus in bar 9 Schubert is repeating his opening violin tune, but he gives it a different colour by making the flute play the same notes an octave higher. Four bars later the colour is changed again by the addition of a clarinet to the tune. After the double bar it is oboe and violins that are doubled. Schubert is continually searching for variety of colour, like any other good orchestrator. In the Minore the interest is mostly given to solo woodwind instruments, as we have seen, but for contrast there are four bars in which flute, oboe, and clarinet all play the tune together.

Note that there is a third section of music called 'Minore II' (page 50–51). If bars 1–32 are called A, and Minore I is called B, and Minore II C, the structure is:

A—B—A—C—A

The repeats in A should only be played the first time round, and on some records they are not played at all.

In the Ballet Music from *Rosamunde* (page 52) the earlier music alternates between strings and woodwind; later there is a charming clarinet solo. Only the principal section is given here; then comes some different music, after which the principal section is played again.

Schubert (1797–1828) wrote his *Rosamunde* music for a very bad play given in Vienna in 1823; the authoress was a rich Hungarian lady called Helmina von Chezy, also notable for providing Weber with one of the worst librettos there is, for his opera *Euryanthe*. *Rosamunde* was a complete failure, and much of Schubert's music for the play was completely forgotten until it was rediscovered in Vienna in the 1860s by two Englishmen, George Grove (who later produced the famous *Dictionary of Music and Musicians*) and Arthur Sullivan, composer of *The Mikado* and other light operas.

B FLAT ENTR'ACTE FROM SCHUBERT'S *ROSAMUNDE*

Minore I.

Maggiore da capo.

Minore II.

Maggiore da capo.

G MAJOR BALLET MUSIC FROM *ROSAMUNDE*

Schubert

AIR FROM BACH'S SUITE III IN D

MOZART'S *EINE KLEINE NACHTMUSIK*

IN the remaining pages of this book you will find the score of a complete work, Mozart's serenade for strings known as *Eine kleine Nachtmusik* (a little night music). Although this uses only four staves, you may find some of it hard to follow at first, especially when the tempo is fast. Concentrate on the first violin part to begin with; this has the tune most of the time. Later see how Mozart accompanies this violin line; the other parts are full of interest. Notice particularly the section in the slow movement with three flats in the key signature; here your eye will have to alternate between the first violin and the cello staves, for these instruments are exchanging ideas in a sort of duet (page 63).

Like so many symphonies, string quartets, and sonatas by Mozart, Beethoven, and Schubert, *Eine kleine Nachtmusik* is in four movements, and some notes on the conventional classical pattern for such music may be of assistance.

First Movement: usually fast, and often the longest and 'biggest' (but Mozart has no ambitions to be 'big' in this charming work). Two sections each repeated (though they may not be on records). The second section is much the longer and ends with a virtual repeat of the first section.

EXPOSITION (the first section). You will notice distinctive tunes beginning at bars 1, 11, 28, and 35.

DEVELOPMENT (the start of the second section). Mozart selects his first and fourth tunes for treatment (bars 56–75).

RECAPITULATION (the completion of the second section). Mozart starts running through his four tunes again at bar 76.

Slow Movement

Here the composer is free to choose from a variety of forms; Mozart picks the rondo, in which one principal tune keeps recurring. Thus the form is:
A (bars 1–16)—B (16–30)—A (30–38)—C (38–50)—A (the rest).

Minuet and Trio: minuets were in two short sections each repeated. To make the movement longer a second minuet, called the Trio, was added. After the trio, the first minuet is played again without repeats.

Finale: usually a rondo or in First Movement form. Mozart's finale is a mixture of the two; it is possible that he wrote the word 'rondo' on the manuscript *before* he wrote the music and then forgot to cross it out, for this piece is scarcely a rondo at all, but more like a 'First Movement'.

SERENADE.
Eine kleine Nachtmusik
für 2 Violinen, Viola, Violoncell und Contrabass
von
W. A. MOZART.
Köch. Verz. No 525.

Componirt am 10. August 1787 in Wien.

Allegro.

Violino I.

Violino II.

Viola.

Violoncello e Basso.

10

20

ROMANZE.

Andante.

MENUETTO.
Allegretto.

EXPRESSION MARKS

Most composers use Italian words for their expression marks, and this is a great convenience, for musicians of all nationalities understand them

SPEED

Presto: very fast (*Prestissimo* is faster still)

Allegro: fast

Allegretto: fairly fast (much the same as *Andantino*)

Andante: at a moderate (or literally *walking*) pace

Lento: slow

Adagio: leisurely, i.e. very slow

Largo: slow and grand (*Larghetto*: not quite so slow as *Largo*)

These words can be qualified by *poco* (slightiy, rather), *moderato* (moderately), *molto* (very) as in *poco allegro, allegro molto*; or by *più* (more), *meno* (less), *ma non troppo* (but not too much). *Sempre più allegro* means 'still faster'. They can also be qualified by the words in the section headed 'Style'. Note also *con moto* (with movement); *più mosso* (faster).

Rit. (ritardando) ⎱ getting slower
Rall. (rallentando) ⎰

String. (stringendo) ⎱ getting faster
Accel. (accelerando) ⎰

These are cancelled by a new tempo indication, or by *a tempo* (in time, i.e. back to the original tempo)

STYLE

Agitato: agitated, restless

Animato: animated

Cantabile: with singing tone

Dolce (or *dol.*): sweetly

Doloroso: sadly

Espressivo (or *espress.*): expressively

Giocoso: cheerfully

Grazioso: gracefully (*Con grazia*: with grace)

Legato (literally *bound, tied*): the notes sound for their full length leading smoothly into each other without gaps; usually indicated by curved lines ('slurs') over or under the notes that are to be bound together; a form of punctuation

Leggiero: lightly

Maestoso: grand, stately

Marcato (or *marc.*): with emphasis, marked; *ben marcato*: well marked

Semplice: simply

Sostenuto: sustained tone

Staccato: the notes sound for less than their due length, with gaps separating them. Usually indicated by dots over or under the notes. The opposite of *legato*

Ten. (tenuto): 'held'; the player lingers on the note but only just perceptibly.

Tranquillo: calmly

Vivace or *Vivo*: lively

Pizzicato: plucked; *arco* (or *co arco*): with the bow

LOUDNESS

ff (*fortissimo*): very loud
f (*forte*): loud
mf (*mezzo-forte*): fairly loud
mp (*mezzo-piano*): fairly soft
sf (*sforzando*): accented
    (sometimes *fz*)
con forza: with force

p (*piano*): soft
pp (*pianissimo*): very soft
cresc. (*crescendo*): getting louder
dim. (*diminuendo*): getting softer
morendo or *smorzando*: dying away
con (*senza*) *sordino*: with (without) mute

CLEFS

All the above notes sound at the same pitch, 'middle C'. Violas use the alto clef, and sometimes trombones do too. Cellos and bassoons sometimes use the tenor clef if their parts lie very high; occasionally trombones use it.

TRANSPOSING INSTRUMENTS

CLARINETS if in B flat, the *sound* is *a tone lower* than you would expect.

if in A, the *sound* is *a tone and a half lower*.

(In classical music clarinets are sometimes in C, and this means that they behave normally.)

HORNS nowadays their parts are always written in F, and this means the *sound* is *a fifth lower* than you would expect.

(In classical music they were often in E flat—sounding a sixth lower—and other keys too.)

TRUMPETS nowadays either in C (when they present no problem) or in B flat or A (when they behave like clarinets). Occasionally in F.

Here are the first three notes of 'Three Blind Mice':

This is how a composer who wanted those sounds would write them for clarinets and for horn in F:

Keys with many sharps or flats are hard for the clarinet, so a composer would choose the clarinet in A.

INDEX OF MUSIC

*(Many excellent recordings exist of this music,
and more appear every month)*